measure and solve™

Activities on
Perimeter and Area
Grades 5-8

Deborah C. Gutman
Carol A. Thornton

Edited by: Jennifer Birmingham
Illustrations by: George Ulrich
Design by: Charles Kulma/Kulma Image & Information Graphics

© 1998 Learning Resources, Inc., Vernon Hills, IL (U. S. A.)
      Learning Resources, Kings Lynn, Norfolk (U. K.)

ISBN: 1-56911-893-0

Printed in the United States of America.

Activities on Perimeter and Area

TABLE OF CONTENTS

Introduction

BORDER PROBLEMS
Perimeter - Circumference

AREAS OF OTHER POLYGONS
Triangles, Trapezoids, and Other Shapes

AREA STRATEGIES
Rectangles - Parallelograms

AREA OF CIRCLES

INTRODUCTION

Measure and Solve: Perimeter and Area provides ideas for engaging grades 5-8 students in problem-centered activities, games, and projects involving perimeter and area. Pattern blocks, color tiles, and tangrams are used throughout the activities of this resource book as students explore ideas and problem-solve. Fast 4! section openers pose shorter, targeted problems to reinforce learning on an ongoing basis. Performance assessments at the end of each section suggest tasks to help teachers profile students' understanding. In addition, applied problems and technology integration are present throughout the book.

This resource may be used as a replacement unit for instruction on measurement. As such, it is divided into four sections that focus on important aspects of measuring perimeter and area:

- *Border Problems: Perimeter and Circumference*
- *Area Strategies: Rectangles and Parallelograms*
- *Areas of Other Polygons: Triangles, Trapezoids, and Other Shapes*
- *Area of Circles*

Each section begins with an introduction page that highlights the key vocabulary and important concepts presented in the pages that follow. Most activities and games assume that students will work collaboratively with a partner and have ready access to available technology and materials necessary to communicate their ideas and findings. Features which recur throughout this book to support these actions are:

A list of materials necessary for the activity.

Activities and suggestions which allow students to explore or review understandings critical to the central TASK of a lesson.

The major problem that guides the activity or project.

Questions which provide opportunity for students to communicate their mathematical understanding.

Questions or ideas posed to prompt an understanding of how or why a concept works.

Calculator or computer alternatives that may be integrated.

Natural extensions of a game or activity.

Section 1:
BORDER PROBLEMS

The following words are used throughout Section 1. An informal definition for these words is listed below. Teachers may want to highlight the words and review their meanings as they appear in the context of each lesson.

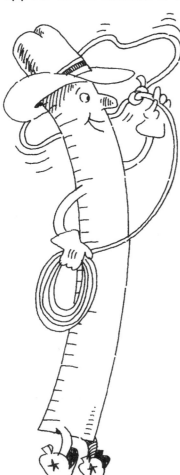

perimeter:	the distance around a shape
area:	the number of square units that fit inside the perimeter of a polygon
circumference:	the perimeter of a circle
concave:	a polygon whose shape turns in on itself
diameter:	the distance across the middle of a circle
equilateral:	having equal sides
hexagon:	a 6-sided polygon
octagon:	an 8-sided polygon
pi (π):	the ratio of the diameter to the circumference of any given circle (3.14)
polygon:	an enclosed, multi-sided figure
radius:	half the distance across a circle (half the diameter)
ratio:	a relationship between two measurements
trapezoid:	a 4-sided polygon in which two sides are parallel
triangle:	a 3-sided polygon

INTRODUCTORY ACTIVITY:

Invite students to measure the perimeter of each pattern block shape. Ask them to identify each shape by name, and label its perimeter in their Math Logs.

Border Problems:
PERIMETER-CIRCUMFERENCE

Fast 4! *(page 8)*

- **Problem 1:** Solutions will vary. Sample: 4 x 4; 6 x 3
- **Problem 2:** Sample solution: 1 x 2 cm and 1 x 4 cm
  Challenge: 9 x 2 and 6 x 3
- **Problem 3:** Possibilities: 1 x 24, 2 x 12, 3 x 8, 4 x 6. 4 x 6 = least perimeter. If you accept decimal or fractional dimensions, there are limitless possibilities.
  Challenge: A 6 x 6 dog run has the least perimeter. Discovery: The rectangle closest to a square shape has the least perimeter.
- **Problem 4:** About 37.68 in.

Perimeter Patterns *(pages 9-10)*

Warm Up: Each hexagon adds 4 to the train's perimeter: 6, 10, 14, 18, 22, 26. Expression: $4n + 2$. As number increases, perimeter increases.

Task: Triangular numbers; perimeter increases by 3 each step. Expression: $3n$. At step 7, the area of the interior shaded region (in triangles) surpasses the perimeter. (P = 21; A = 25) At step 75, perimeter = 225; shaded area = 1444 triangles.

Next Steps: $P = 4n$; shaded area sequence {0, 0, 0, 1, 3, 6, 10, 15, 21, 28...} At step 13, the shaded area (55) surpasses the perimeter (52).

Perimeter Puzzles *(pages 11-12)*

There are multiple solutions for each puzzle and each perimeter.
All perimeters (6 minimum to 18 maximum) for Puzzle #6 are shown below.

<u>Puzzle #6</u>

P=6   P=7   P=8   P=9   P=10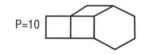

P=11 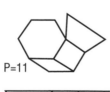   P=12    P=13 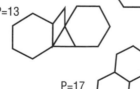   P=14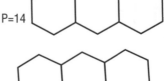

P=15 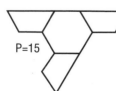   P=16    P=17   P=18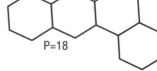

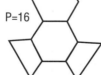

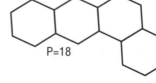

Teaching Notes/Answer Key

Custom Fences (pages 14-15)

Task: Most students decide on Fence Package D for Mr. Jones, C for Mr. Barcia, A for Ms. Yu, E for Mr. Robinson, and B for Ms. Greene, but accept other logical decisions. The work below represents a sample response and set of drawings for designing the different packages.

Mr. Jones

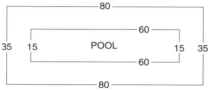

Package D works best for Mr. Jones because he wanted some grass between the fence and the pool. He has 10 feet between the pool and the fence. Fence length is 230 ft.

Mr. Robinson

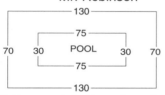

Mr. Robinson needs the biggest fence of all. He had the largest amount of land, a poolhouse. and a deck to fence in, and money was no problem. Package E fits his needs best. Fence length is 400 ft.

Ms. Yu

The fence is lined up exactly with the deck as Ms. Yu asked. She didn't specify any amount of money, but she got the most fence for her money with Package A. Fence length is 80 ft.

Ms. Greene

Ms. Green's fence has only 5 feet on 3 sides (between the pool and the fence), and 6 feet on other side. She got the most fence for her money with Package B. Fence length is 120 ft.

Mr. Barcia

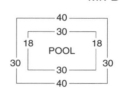

Mr. Barcia has specific space limitations so there was only one logial way to position his fence. Package C works best for this reason. Fence length is 140 ft.

Frisbee Go-Round (page 17)

Warm Up: A little more than 3 times; diameter; circumferences; yes

Communication: $C/d \approx 3.14$; 3.14; $d = C/\pi$; $C = \pi \times d$

Circumference Dominoes (page 18) is self-checking.

Performance Assessments (pages 19-20)

Assessment 1: 4 (36) = 144 cm; 2 (21.5) + 2 (50) = 143 cm;
2 (31) + 2 (42) = 146 cm; 2 (23)(3.14) ≈ 144.44 cm. Rank: B, A, D, C

Assessment 2: Situation A
• 2 (2) + 2 (5) = 14 km • 4 (4) = 16 km • 5 (3.14) ≈ 15.7 km
The square path is the longest.

Situation B
• 2 (0.5) + 2 (1.2) = 3.4 km • (0.8)4 = 3.2 km • (1.1)π ≈ 3.45 km

The longest jog would be the circular path around the statue.

Assessment 3: 3 (17) = 51 units

FAST 4!

1
Draw a rectangle that has the same unit perimeter and area.
Hint: Use color tiles or graph paper.

Challenge: Find three more rectangles whose individual unit perimeter and area are the same.

2
Draw two rectangles that differ in area by 2 square centimeters, and differ in perimeter by 4 centimeters.

Challenge: Find two rectangles that have the same area, but differ in perimeter by 4 units.

3
One family decided to fence in 24 square yards of backyard space so their new puppy could have a special place to romp and roam. They decided the space should be rectangular.

What different dimensions are possible?

Which design requires the least fencing?
(Hint: Use color tiles to help decide, letting 1 tile represent 1 square yard.)

Challenge: The children in this family would like a bigger space for the dog (36 square yards). Sketch all possible designs for this space, and decide which requires the least amount of fencing.

4
You may have learned that you can find the age of a tree by counting the tree's rings. A certain tree, struck by lightning, was found to have 120 rings and a 6-inch radius. How big around was this 120-year-old tree?

What patterns can you find?

PERIMETER PATTERNS

Generous supply of pattern blocks for each pair of students, Math Log, and a writing utensil.

With a partner, make a train of six hexagons. As you add each hexagon, record the perimeter of the train in the chart below. Then answer the questions that follow in your Math Log.

# of Hexagons	Perimeter of Train
1	
2	
3	
4	
5	
6	
n	

- How did the train's perimeter grow with each additional hexagon?
- Explain why the perimeter increased as it did (not by 5 units each time).
- Write a mathematical expression for the perimeter of any hexagon train. Use **n** to stand for the number of hexagons.   **P** = _____

TASK

  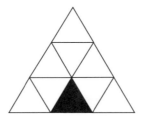 

Use pattern blocks to create each of the first three steps in the triangular number pattern. Sketch the first four steps in your Math Log.

- What type of pattern is this?
- Describe how the perimeter increases in each step.
  Extra for Experts: Write a mathematical expression to describe this increase.
- What pattern do you notice with the shaded interior regions?

BORDER PROBLEMS

What patterns can you find?

MORE PERIMETER PATTERNS

PROBLEM: At which step will the area of the shaded interior region surpass the perimeter of the triangle?

	Step									
	1	2	3	4	5	6	7	8	9	n
Perimeter of Triangle										
Area of Shaded Triangle (in blocks)										

COMMUNICATE

Compare your results with the class.

- How did you go about counting the perimeter for each step?
- What is the pattern for the area of the shaded region?
- Discuss how you could know the perimeter and shaded area of any triangle in this pattern. Use your pattern to find out the perimeter and shaded area of the triangle in Step 75!

Step 75: Perimeter =_____   Shaded Area =_____

NEXT STEPS...

Look carefully at this step pattern:

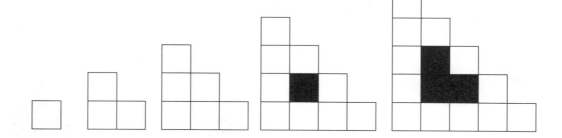

Create the first three steps in the pattern with color tiles. Sketch the first five steps in your Math Log.

- Figure out the pattern for perimeter.
- Describe the pattern for the shaded interior tiles.

Now determine when the shaded interior area of each figure will be greater than its perimeter. Record your work in your Math Log.

BORDER PROBLEMS

Perimeters of Polygons

Can you solve and create perimeter puzzles?

PERIMETER PUZZLES

Pattern blocks for each pair of students.

WARM-UP

Using the clues, create as many solutions to the puzzle as you can.
Sketch each solution in your Math Log.

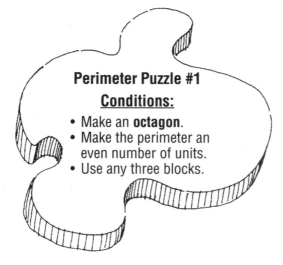

Perimeter Puzzle #1

Conditions:

• Make an **octagon**.
• Make the perimeter an even number of units.
• Use any three blocks.

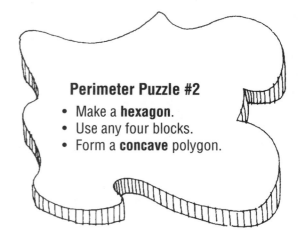

Perimeter Puzzle #2

• Make a **hexagon**.
• Use any four blocks.
• Form a **concave** polygon.

TASK

How many possible perimeters can you find for these puzzle conditions?

MEASUREMENT NOTE:

The base of the green triangle is equal to 1 unit of perimeter. Each triangle equals one unit of area.

COMMUNICATE

Sketch the polygons and record your solutions in a table in your Math Log:

Use the following headings:
• **Perimeter** • **Sketch**

Share at least one solution for each possible perimeter as a class.

• What would be the least possible perimeter? Why?

• What is the greatest perimeter you could find?
How would you know when you have reached the maximum perimeter limit?

• Why do you think you were able to find more than one possibility for the same perimeter?

BORDER PROBLEMS

MORE PERIMETER PUZZLES

TASK

Sketch solutions for each of these perimeter puzzles in your Math Log. How many unique solutions can you find?

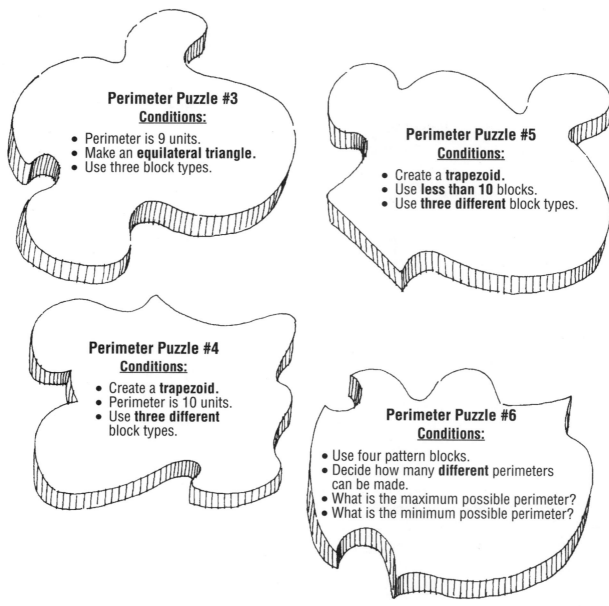

Perimeter Puzzle #3
Conditions:
- Perimeter is 9 units.
- Make an **equilateral triangle.**
- Use three block types.

Perimeter Puzzle #5
Conditions:
- Create a **trapezoid.**
- Use **less than 10** blocks.
- Use **three different** block types.

Perimeter Puzzle #4
Conditions:
- Create a **trapezoid.**
- Perimeter is 10 units.
- Use **three different** block types.

Perimeter Puzzle #6
Conditions:
- Use four pattern blocks.
- Decide how many **different** perimeters can be made.
- What is the maximum possible perimeter?
- What is the minimum possible perimeter?

NEXT STEPS...

Write your own perimeter puzzles for classmates to solve.

A Partner Game

THE GRID GAME

For each pair of students: grid paper (p.62), 2 (1-6) dice, two different colored pencils.

Objective: Given an area, create polygons with the greatest possible perimeter on the grid.

GAME RULES

- Players choose a color.
- Each player rolls a die. Highest roll goes first.
- In turn, roll and add the numbers on two dice.
- On the grid, outline and color a polygon having an area equal to the dice sum. Polygons cannot overlap on the grid.
- Count and record the polygon's perimeter.
- The game ends when both players have rolled areas that cannot fit on the grid.
- The player with the higher total perimeter wins!

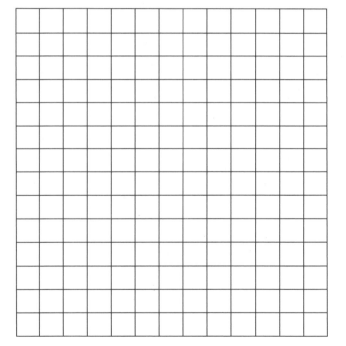

Player A: Perimeter	Player B: Perimeter

- Which strategies could be used to get the greatest possible perimeter?
- Explain the strategies you used to block your opponent.
- How can a given area have different possible perimeters?

COMMUNICATE

BORDER PROBLEMS

Applied Problems:
Perimeter of Rectangles

What is the best fence package for each customer?

CUSTOM FENCES

Calculator and writing utensil for each pair of students.

WARM-UP

You are working for a fence company. By law, all swimming pools must be fenced in. Your first assignment is to design **all possible** fences to enclose **rectangular** swimming pools and patios covering areas of **60 square units.** (You can assume that these fences will surround all four sides of the pool property.) Sketch your fence designs in your Math Log or on grid paper. Record the fence length in each design.

COMMUNICATE

- How many different designs did you find?
- How did you know when you had **all** possibilities?

Compare the perimeters for all designs:
- Which design would be most economical to sell?
- Which designs would sell the best? Why?
- What factors would you consider when pricing each of your designs?

TASK

Your boss was very pleased with your first assignment! Now he wants you to match five unique situations to standard fence packages that your company sells. *Which fence package would you sell to each customer?*
Complete the table on page 15 as you work.

NEXT STEPS...

- Survey friends, asking them to describe their "dream pool."
- Pick an interesting request, and decide how you might fill the order using a standard fence kit while meeting the conditions described on page 15.
- Be prepared to explain the strategy you used to best meet the customer's desires!

What is the best fence package for each customer?

MORE CUSTOM FENCES

TASK

You are working for a fence company. Your job is to match a standard fence kit to your customers' unique needs. Your five customers are enclosing swimming pool properties as required by law. Which fence package would you sell to each customer?

CONDITIONS

- Use the pool dimensions and property information listed below.
- All fenced areas are rectangular, completely enclose the pool, and are measured in feet.
- A customer's dimensions may not match a standard kit exactly. Use your judgement to make the best fit. Use the least expensive fence package possible to save your customers' money.
- Sketch the dimensions of each customer's property to be fenced on grid paper.
- Organize your information in the table below.

Customer	Pool Dimensions (feet)	Conditions / Yard Data	Fenced Area Dimensions (ft.)	Fencing Length Needed (ft.)	Fence Package
Mr. Jones	Lap Pool 15 x 60	Wants some grass between pool and fence. Wants most area for the money.			
Ms. Yu	Kidney shaped 8 x 10	A 15-foot deck will form one side of the area to be fenced. Wants the deck to line up exactly with the fence.			
Mr. Robinson	Rectangular 30 x 75	Located on a 5-acre property. Wants room for a 10 by 12 bath house and a 20-foot deck inside the fence. Size and expense are no problem.			
Ms. Greene	Rectangular 15 x 24	Wants smallest perimeter of fencing possible. Needs a minimum of five feet of clearance on each side of the pool.			
Mr. Barcia	Rectangular 18 x 30	Space limitations. Backyard is 30 x 40. Wants most area possible.			

COMMUNICATE

Write a journal entry to your boss at the fence company explaining your reasoning for each customer's package.

FENCE PACKAGES

A	80 foot fence	$800
B	120 foot fence	$1000
C	180 foot fence	$1500
D	240 foot fence	$2000
E	500 foot fence	$4000

A Partner Game

THE "MIN WINS" GAME

For each pair of students: one die (0-5 or 1-6), pattern blocks, score sheet or Math Log, and a writing utensil.

Objective: See how perimeter changes as the shape and area of a polygon changes.

GAME RULES

- Each player rolls a die. Lowest roll goes first.
- In turn, roll the die to determine the number of pattern blocks to use.
- Use the pattern blocks to form a polygon with the *least possible* perimeter.
- Each turn, add new pattern blocks to your polygon.
- Record how your perimeter changes on the score sheet.
- After 10 rounds, the polygon with the least perimeter wins!

PLAYER 1

Turn	# Blocks Used	Total Perimeter

PLAYER 2

Turn	# Blocks Used	Total Perimeter

COMMUNICATE

- Explain the strategies you used to keep your perimeter as low as possible.
- Which blocks helped the most?
- What happened to the perimeter as the area increased?

NEXT STEPS...

- Play the same game with the **maximum** perimeter as the goal.

What is the relationship between the circumference and diameter of a circle?

FRISBEE GO-ROUND

For each pair of students: large frisbee-sized disc, non-stretchy string, rulers marked in millimeters, 5 assorted lids or other circular objects, calculator, and a writing utensil.

WARM-UP

Using only unmarked string, try to figure out how many times the distance across the large disc will fit around its perimeter. Share your findings with the class.

- Do you think the relationship you found is true for every circle in the universe?
- What do we call the **"distance across the disc"** and its **perimeter?**

TASK

What do you notice about the relationship between the circumference and diameter of a set of circles?

- Working with a partner, measure the circumference and diameter of each of your lids to the nearest millimeter. Record your results in the chart below.
- Use a calculator to figure out how many diameters fit into each circumference. Round these numbers to the hundredths place.

Object Measured	Circumference	Diameter	Circumference Diameter
			Average =

COMMUNICATE

- Answer the question posed in the TASK. Compare with other groups.
- The relationship between circumference and diameter is a called **a ratio.** Any diameter fits into the circumference **a little more than** _____ times.
- A more exact number for this ratio is called **pi (π)**, which is rounded for common use to _____.
- How could we figure out the diameter if we know the circumference?
- How could we get the circumference if we know the diameter?

BORDER PROBLEMS

Circumference of a Circle

A Partner Game

CIRCUMFERENCE DOMINOES

> Remember... C = π x d   (Use 3.14 for π)

GAME RULES
- Cut out the game cards and deal 5 to each player.
- Place one card face up between players to start a domino train.
- Shuffle the remaining cards and place them face down in a pile.
- In turn, draw a card.
- Either add to the domino train or add the new card to your hand.
- Players may use a calculator to help solve the problem.
- First player to run out of cards wins!

Cut out these **GAME CARDS** on the dotted lines:

r = 8 cm C ≈	6 inches	d = 15' C ≈ _____	≈ 37.68"	12 in. C ≈ _____	7.5 inches	C ≈ 31.4 cm diameter = _____	≈ 39.25 inches
d = 12.5 in. C ≈ _____	≈ 25.12 ft.	About how many cm around is a hose with a 2.5 cm diameter?	10 cm	Give the length of lace needed to frame a photo having a 10-inch diameter.	47.1 cm	C ≈ 12.56 cm radius = _____	6.5 ft.
C ≈ 40.821 cm radius = _____	≈ 31.4 in.	C ≈ 69.1 cm radius = _____	2.5 ft.	4' C ≈ _____	≈ 21.98 inches	Find the circumference of a tree stand needed to fit a tree having a radius of 3.5".	≈ 50.24 cm
What is the diameter of a snake having a circumference of about 4.7 inches?	9 inches	What would be the diameter of a tree stump having a circumference of about 28.26 inches?	≈ 18.84 inches	What is the diameter of a paper plate that is about 25.12 inches around?	2 cm	C ≈ 23.55" diameter = _____	7.85 cm
15 cm C ≈ _____	8 inches	Give the diameter of an elephant's snout having about an 18.84-inch circumference.	11 cm	3 cm C ≈ _____	≈ 47.1 ft.	C ≈ 15.7 feet radius = _____	≈ 1.5 inches

PERFORMANCE ASSESSMENTS 1-2

ASSESSMENT 1: Rubric 4   3   2   1   0

Calculate the perimeter of each figure. In your Math Log, arrange them in order from least to greatest. Show or explain your calculations.

36 cm

A

36 cm

50 cm

B

21.5 cm

31 cm

C

42 cm

D

23 cm

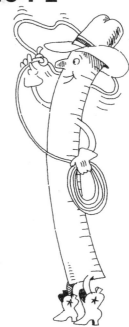

ASSESSMENT 2: Rubric 4   3   2   1   0

Respond to either situation below.

Situation A

You and your friend want to bicycle together. There are three paths in your neighborhood:
- One goes around a rectangular block which is 2 km x 5 km.
- The second goes around a 4 km square park.
- The third is a circular path around a statue. The diameter of the circular path is 5 km.

Which path would give you the **longest** ride? Show your calculations, and explain your reasoning in your Math Log.

Situation B

You and your friend want to jog together. There are three paths in your neighborhood:
- One goes around a rectangular block which is 0.5 km x 1.2 km.
- The second goes around a 0.8 km square park.
- The third is a circular path around a statue. The diameter of the circular path is 1.1 km.

Which path would give you the **longest** jog? Show your calculations, and explain your reasoning in your Math Log.

RUBRIC for these assessments:

4 Fully successful with the perimeter task.
Makes correct calculations. Work is shown and appropriately labeled/justified.

3 Substantially successful with the perimeter task. Calculation strategies show correct concept of perimeter/circumference. May contain minor calculation/labeling errors.

2 Partially successful with the perimeter task. Calculation strategies show correct concept for *either* rectangles or circles. Work may not be appropriately labeled/justified.

1 Engaged in the perimeter task. Reteaching of perimeter/circumference is necessary for both rectangles and circles.

0 No attempt or non-mathematical response.

BORDER PROBLEMS

PERFORMANCE ASSESSMENT 3

Complete the task below.

ASSESSMENT 3:   Rubric 4   3   2   1   0

Calculate the perimeter of this figure if the base of the triangle equals 3 units. Show your work and explain your strategy in your Math Log. Label your answer.

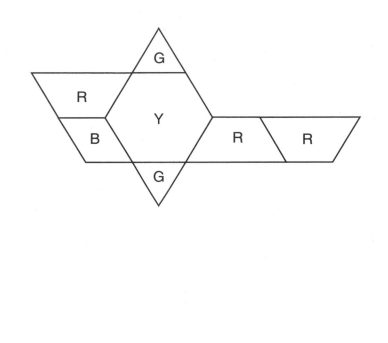

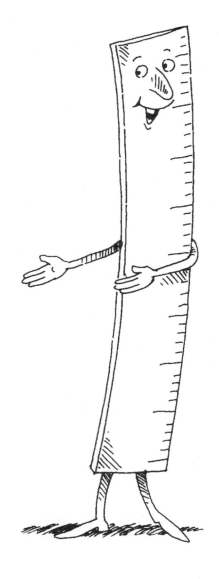

RUBRIC

4 Fully successful with the perimeter task.
   • Calculations show correctly labeled perimeter expression.

3 Substantially successful with the perimeter task.
   • For the most part work is correct, though it may need minor error correction(s).
   • Minor errors may include an addition or multiplication error and/or incorrect labeling.

2 Partially successful with the perimeter task.
   • Some attempt was made at finding perimeters; however, reteaching is necessary.
   • Perimeter attempts may include labeled sides and/or the addition of some sides.
   • A correct perimeter response is not attained.

1 Engaged in the perimeter task.

0 No attempt or non-mathematical response.

Section 2:
AREA STRATEGIES

The following words are used throughout Section 2. An informal definition for these words is listed below. Teachers may want to highlight the words and review their meanings as they appear in the context of each lesson.

area: the number of square units needed to cover a surface

conventional: the most-often used style or expression

parallelogram: a 4-sided polygon with opposite sides parallel

perfect: a polygon with equal sides and equal angles

prime: a number whose only factors are 1 and itself

quadrilateral: a 4-sided polygon

rhombus: a quadrilateral having four equal sides

scale drawing: a drawing in which each measured distance is proportional to the actual distance on the object represented

square units: the conventional unit of measurement for area

square: a perfect 4-sided parallelogram

trapezoid: a quadrilateral with two, and only two, parallel sides

Area Strategies:
RECTANGLES-PARALLELOGRAMS

Fast 4! *(page 24)*

- **Problem 1:** Lake A covers more area. Possible strategies include counting the squares, using an "averaging" method for the partially covered squares, inscribing the lakes in rectangles, and folding the lakes on top of each other, comparing the difference.

- **Problem 2:** Rectangles having perimeter = 24; 1 x 11; 2 x 10; 3 x 9; 4 x 8; 5 x 7; 6 x 6.

- **Problem 3:** Either slice would contain 18 square inches of pizza.

- **Problem 4:** C (19 square units); B (20 square units); A (25 square units).

Rolling Rectangles Game *(page 27)*

6 is the only perfect number using 1-6 dice. 28 would be possible with larger-numbered dice.

Parallel Play *(page 28)*
Warm-Up:

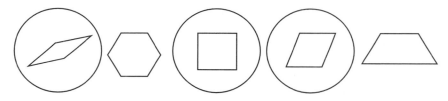

Max/Min Problems *(page 29)*

Task: <u>Sandbox Problem</u>: A 106.25 square would have the largest area.

<u>Garden Problem</u>: A 30 x 30 garden would yield 900 square feet of space vs. 800 for the 40 x 20.

<u>Quilt Problem</u>: A 59 x 59 inch quilt would have the largest area = 3481 sq.in. Minimum binding needed for an 864 sq. inch rug ≈ 117.58 in.

Dream House Design *(page 30)*

Task: Sample student responses and Rubric on p. 23.

Performance Assessments *(page 31)*

- **Assessment 1:** Least to greatest by area: D, B, A, C.

- **Assessment 2:** The 6 x 5 would hold the largest photo and would probably sell the best, as it is closest to commercial photo sizes. The maximum area would be obtained from the rectangle most closely resembling a square. The minimum photo area would be from a 1 x 10 frame, which is too narrow for standard photos.

scale shown: 1/4" = 2 ft.

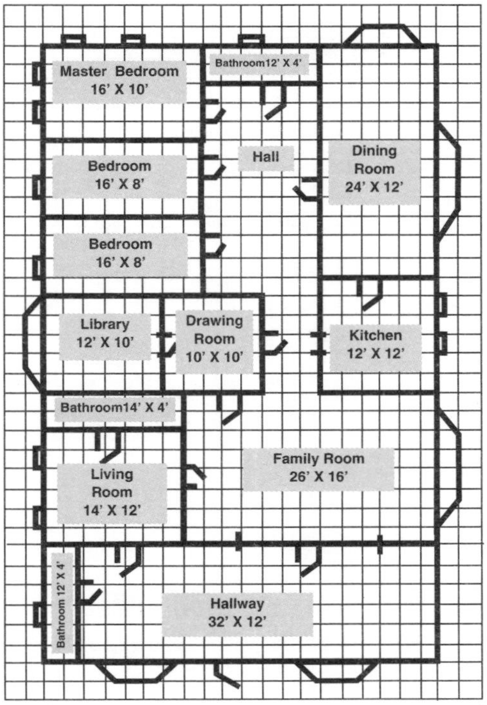

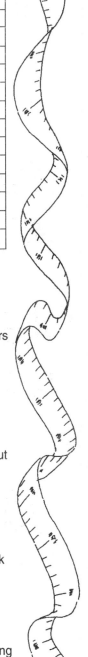

Dream House Spreadsheet (Flooring Costs)

Room	Length	Width	Area sq. ft.	Unit Price	Cost
Hallway 1	32	12	324	5.99	1,940.76
Bathroom 1	4	12	48	2.99	143.52
Family Room	26	16	416	3.99	1,659.84
Living Room	14	12	168	5.99	1,006.32
Bathroom 2	14	4	56	2.99	167.44
Library	12	10	120	5.99	718.80
Drawing Rm.	10	10	100	5.99	599.00
Hallway 2	-	-	324	5.99	1940.28
Kitchen	12	12	144	2.99	430.56
Dining Rm.	12	24	288	5.99	1,725.12
Bathroom 3	12	4	48	2.99	143.52
Bedroom 1	16	8	128	5.99	766.72
Bedroom 2	16	8	128	5.99	766.72
Master Bed.	16	10	160	5.99	958.40
				Subtotal	112,967.00
				Tax (5%)	648.35
				Grand Total	13,615.35

Scale drawing floor plan labels:
- Master Bedroom 16' X 10'
- Bathroom 12' X 4'
- Bedroom 16' X 8'
- Hall
- Dining Room 24' X 12'
- Bedroom 16' X 8'
- Library 12' X 10'
- Drawing Room 10' X 10'
- Kitchen 12' X 12'
- Bathroom 14' X 4'
- Living Room 14' X 12'
- Family Room 26' X 16'
- Bathroom 12' X 4'
- Hallway 32' X 12'

Dream House Rubric

4 Fully Accomplished the Task

Scale Drawing
- Accurate use of scale
- Dimensions are correct
- House fits on the given lot
- Minimum of six rooms

Spreadsheet
- Formulas used to accurately calculate area, costs, tax and totals

Persuasive Advertisement
- Topic Sentence
- Minimum of three details/ sentences/ arguments
- Conclusion

3 Substantially Accomplished the Task

Scale Drawing
- Accurate use of scale
- May have minor area errors in one or two rooms
- House may be too wide for the lot
- Minimum of 6 rooms

Spreadsheet
- Area calculations and flooring costs have two or less errors
- Tax and totals are correct for given subtotal

Advertisement
- Contains topic sentence, details and conclusion, but may need more detail

2 Partially Accomplished the Task

Scale Drawing
- Attempted to use scale, errors may be present
- Calculation errors found in four or fewer rooms
- House may not comply to all conditions

Spreadsheet
- Length, width, areas and flooring costs are entered, but may be incorrect for four or fewer rooms
- Tax and totals may be incorrect

Advertisement
- An ad is included in the task
- Topic sentence may be weak
- May need more details
- Conclusion may be missing

1 Engaged in the Task
- Scale drawing and spreadsheet have been attempted
- Persuasive ad may be missing

0 No Attempt

FAST 4!

1 Estimate which lake covers more area.
Explain your choice and your strategy.

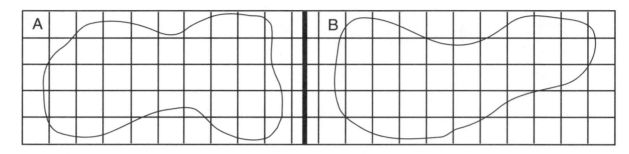

2 Sketch all possible rectangles having a perimeter of 24 units but different areas.
Use color tiles to plan your sketches. Share your findings with a partner.

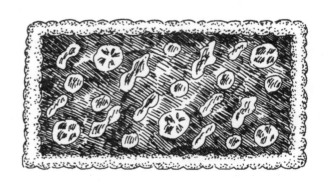

3 Would you rather have one of 16 "fair share" slices of a 12" x 24" rectangular pizza or one of 8 "fair share" slices of an 8" x 18" rectangular pizza? Explain your reasoning.

4 Rank these shapes from least to greatest by area. Explain your reasoning.

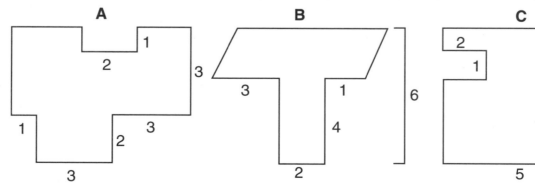

How do you find the area of any rectangle?

EXPLORING RECTANGLES

For each pair of students: color tiles, envelope, cereal box panel, ruler, calculator, grid paper (p. 62), and a writing utensil.

WARM-UP

With a partner, use color tiles to estimate the area of three rectangles in square inches: an envelope, the front panel of a cereal box, and the surface of your desk.

COMMUNICATE

- How many 1-inch squares did you count along the length of each rectangle?
- How many rows of 1-inch square tiles filled each rectangle?
- Explain your area estimate for each rectangle.
- How did you estimate when the tiles did not fit exactly?
- Explain a strategy for estimating your desk's surface area without having enough tiles to cover it.

NEXT STEPS...

Using a ruler, measure the base and height of each rectangular object to the nearest quarter of an inch.
Use your strategy from above and a calculator to compute the area of each object. *How reasonable were your estimates?*

REMINDER:
Area is the number of units needed to cover a surface.
Square units are conventional.

TASK

The classroom seems too small for all the furniture and people to move around comfortably. The teacher wants you to design an arrangement of the furniture in this classroom to save the most floor space. The class will then vote on the design that would work the best. Use the activity sheet on p. 26 for your work.

- On grid paper, sketch the floor layout of the classroom described on page 26.
- Prepare a persuasive speech explaining why your plan is the best.
- Share your design and its advantages in an oral presentation.

Use Geometer's Sketchpad to sketch and measure the areas of the classroom furniture. Then use the Draw option of Claris Works or another similar program to show your space-saving floor plan.

AREA STRATEGIES

How much floor space can you create in this classroom?

A SPACE-SAVING DESIGN

Conditions

- The classroom is 28 x 40 feet. Mark the door, board, and windows together with your classmates.
- Fit in 28 desks, each 2 x 2.5 feet; a 3 x 5 foot teacher's desk, two 1 x 6 foot bookcases, and three 4 x 3 foot computer desks.
- All students must be able to see the board.
- Sketch your floor plan on grid paper; provide a key to the scale used. (e.g. 1 square unit = 1 sq. ft. or 1 square unit = 2 sq. ft.) Use arrows to show walkways for students and teachers.
- Calculate the floor area covered by all of the furniture.
- Calculate the open floor space that is left.

FLOOR PLAN FOR MY SPACE-SAVING DESIGN

(Staple your grid paper here
so it will unfold to show your floor plan.)

- How much total area is covered by furniture in your floor plan?
- How much open area is provided by your plan?
- What are the three best features of your floor plan? Why do these features "stand out"?

COMMUNICATE

- Talk about these facts in an oral presentation about your space-saving design.

An Area Game of Categories

ROLLING RECTANGLES GAME

For each pair of students: grid paper (p. 62), two dice, one score chart (below), and a writing utensil.

Objective: Find rectangle areas that meet given conditions.

GAME RULES

- Roll the dice. Those numbers reflect dimensions of a rectangle.
- Sketch the rectangle on grid paper: label the dimensions, area, and perimeter.
- Enter the area of your rectangle as your "score" in one of the ten boxes below.
- If the area will not fit a category, enter it in CHANCE (if available) or enter a zero score in the box of your choice.
- Alternate rolls for ten turns. If you fit all categories, score 10 extra bonus points.
- Total your column. Highest score wins!

SCORE CHART

Category	Player 1: Score	Player 2: Score
1) Area (A) = Perimeter (P)		
2) Area = Even number		
3) P – A = 4 or A – P = 4		
4) Area = Perfect square		
5) Perimeter > Area		
6) CHANCE		
7) Area = Odd number		
8) Area = Prime number		
9) Area = Perfect Number		
10) Area > Perimeter		
Bonus Points (10):		
GRAND TOTAL		

- Which categories were hardest and easiest to roll? Why?
- How many different areas with perfect number dimensions can be rolled?

COMMUNICATE

AREA STRATEGIES

What can you discover about the area of a parallelogram?

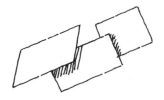

PARALLEL PLAY

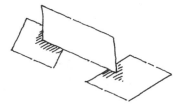

For each pair of students: pattern blocks, one index card, a metric ruler, and a writing utensil.

WARM-UP

Select all pattern blocks that are parallelograms. How many could you find?

TASK

Make a parallelogram out of **more than three** pattern blocks. Outline the parallelogram on an index card, and cut it out.

- Without using a ruler, estimate the area of the parallelogram in square centimeters.

- Write your estimate on the index card figure. Compare estimates and strategies. Do they seem reasonable? Which estimates would you question?

- Making only one cut, move one piece of the parallelogram to form a rectangle.

- Use a ruler to measure the base and height of the new rectangle.

- Calculate the exact area of the index card rectangle to the nearest tenth. Record the area on the card.

- Reassemble the figure as a parallelogram.

COMMUNICATE

- Was your parallelogram area estimate reasonable?

- Show the base and height of both the index card rectangle and parallelogram. What do you notice?

- Explain to a friend your rule for calculating the area of any parallelogram. Write it down for him or her as well.

NEXT STEPS...

Use what you have learned to calculate the **areas** in square centimeters **of all pattern blocks.**

AREA STRATEGIES

How can you explore maximum and minimum measurements in different ways?

MAX/MIN PROBLEMS

For each pair of students: 36-inch piece of string, ruler, graphing calculator, and writing utensil.

With a partner, experiment with a 36-inch piece of string. Shape the string into a parallelogram, a square, and a non-squared rectangle. Calculate the area of each shape by measuring the base and height. Determine which shape has the maximum and minimum area for the same perimeter.

- Use the list function of your graphing calculator to solve the following max/min area problems.
- Discuss your findings. Is there a pattern?

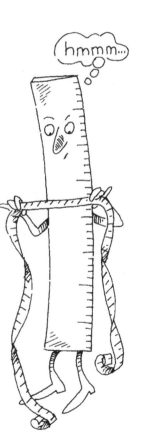

Calculator Steps

Clear any pre-existing data. Use the manual from your calculator to help you enter possible lengths in LIST 1 and corresponding widths in LIST 2. In LIST 3, enter an equation for the perimeter and in LIST 4, enter the area.

Sandbox Problem

Your weekend project is to build a sandbox for your little sister in the back yard. Your parents have given you enough wood for a perimeter of 425 cm. Figure out the dimensions of the largest possible sand box that you can build with the given wood.

Garden Problem

This spring, your family wants to plant a rectangular vegetable garden. The only free spaces in your yard that get enough sun are a 40 x 20 foot section and a 30-foot square. You have purchased the 120 feet of fencing for your garden. Before digging, you need to figure out which location would give you the largest area. Explain your decision.

Binding Problems

Your class is making a quilt and a rug for a service project. Binding is used around the perimeter of both the quilt and the rug.
- Look at several possible quilts; find the greatest area you could make with 236 inches of binding. What would be the smallest useful size?
- Keeping in mind a useful-shaped rug size, what's the minimum amount of binding needed for a rug having an area of 864 sq. inches?

Write a Math Log entry on one of the following:

- How can a calculator help you compare possible dimensions for an area/perimeter problem?
- Describe a situation in which finding the maximum possible area for a given perimeter would be necessary.

COMMUNICATE

© Learning Resources, Inc.

AREA STRATEGIES

Can you sketch a floor plan and cost out the flooring?

DREAM HOUSE DESIGN

For each pair or group of students:
grid paper (p. 62) and a writing utensil.

TASK

Design a **scale drawing** of a house with the
following conditions:

- Use grid paper;
  scale: $\frac{1}{4}$ inch square = 2 square feet.
- The lot is 50 x 120 feet.
  County ordinances require 5 feet of
  clearance from each lot line.
- Sketch a minimum of six rooms.
- Label the dimensions of each room.
  (Example: 9' x 12')
- Choose tile, carpeting, marble, or hardwood for the flooring:

Flooring Costs (per square foot)			
Tile	$2.99	Marble	$4.99
Carpeting	$3.99	Hardwood	$5.99

Track the cost of the flooring choices in the table below.

ACTIVITY SHEET:

	Room	Length	Width	Area (square feet)	Flooring Choice	Unit Cost	Flooring Cost
1							
2							
3							
4							
5							
6							
7							
8							
9							
10							
11							
12							

COMMUNICATE

- Write a persuasive advertisement explaining the benefits of your house design.

PERFORMANCE ASSESSMENTS 1-2

ASSESSMENT 1: Rubric 4   3   2   1   0

Measure and calculate the area of each figure and arrange them in order from least to greatest. Show or explain your calculations.

A   B   C   D

ASSESSMENT 2: Rubric 4   3   2   1   0

Sketch and label the dimensions of two picture frames having a 22 in. border that will hold the largest and smallest possible photos.
Dimensions must be **whole** numbers and need not be drawn to scale.
Explain which frame would sell better.

Sketches: Label length, width and area.

Which frame would sell better? Why?

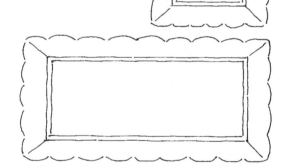

RUBRIC

4 Fully succeeded in finding areas of rectangles, squares, and parallelograms and demonstrating area/perimeter relationships.

3 Substantially succeeded in finding areas. Minor calculation errors only.
- Area calculations are correct for at least three figures in Assessment 1.
- Dimensions are correct for Assessment 2; calculations may have a minor error.

2 Partially succeeded in finding areas.
- Calculations are correct for two of the figures in Assessment 1.
- Succeeded in finding different possibilities for areas of rectangles sharing the same perimeter.

1 Engaged in the task.
- Calculations were attempted for Assessment 1.
- Frames were drawn and dimensions labeled in Assessment 2.

0 No attempt or non-mathematical response.

AREA STRATEGIES

Section 3:
AREAS OF OTHER POLYGONS

The following words are used throughout Section 3. An informal definition for these words is listed below. Teachers may want to highlight the words and review their meanings as they appear in the context of each lesson.

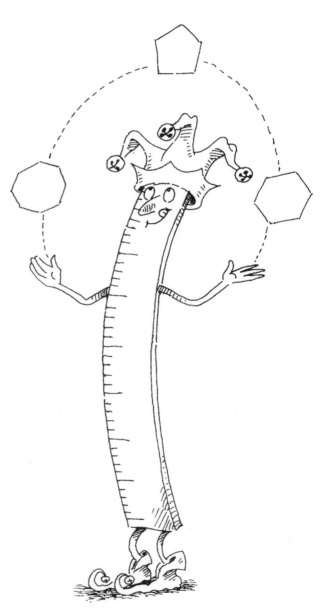

coordinate plane:	grid with y and x axis on which points lie
regular polygon:	a polygon with equal sides and equal angles
irregular polygon:	a polygon with unequal sides
pentagon:	a 5-sided polygon
hexagon:	a 6-sided polygon
heptagon:	a 7-sided polygon
octagon:	an 8-sided polygon
nonagon:	a 9-sided polygon
decagon:	a 10-sided polygon

Areas of Other Polygons:
TRIANGLES, TRAPEZOIDS, AND OTHER POLYGONS

Fast 4! *(page 35)*

- **Problem 1:** Possible estimate: (.5)(210 + 520)(310) = 113,150 sq. miles.
  Atlas estimate is 109,806 sq. miles.
- **Problem 2:** Trapezoid; area = (.5)(5 + 7)(3) = 18 sq. units
- **Problem 3:** x = 4; small triangle: A = 6 cm^2, P = 12 cm; large triangle: A = 60 cm^2,
  P = 40 cm
- **Problem 4:** 16 cm^2

Pattern Block Polygons *(page 40)*

Task: Areas will vary depending on the number and type of pattern blocks used. Interior angle sums: pentagon 540°, hexagon 720°, heptagon 900°, octagon 1080°, nonagon 1260°, decagon 1440°. Pattern: adding a side to a polygon increases its interior angle sum by 180° (line).

Area Bingo *(page 41)*

Enlarge the cards on the copier; glue to construction paper and laminate (optional). Cut out and store in small plastic bags. Note that shapes on the game cards are not drawn to scale. Both partners need to agree upon the solution, as they both have every answer on their boards; only the positions are different.

Trapezoid Patterns *(page 43)*

Next Steps:

```
R                  R                  R
R B                R B                R B
R B B              R B B              R B B
R B B B            R B B B            R B B B
R B B B B          R B B B B          R B B B B
R B B B B B        R B B B B B        R B B B B B
                   R B B B B B B      R B B B B B B
                                      R B B B B B B B
```

Communication: Number of blues will double number of reds at Step 5; triple at Step 7. Blue area = red area at Step 4. Blue area > red area at Step 5. Blue area = double red area at Step 7; blue area triples red at Step 10.

	# red	# blue	ratio R:B		# red	# blue	ratio R:B
1.	1	0	1:0	7.	7	21	1:2
2.	2	1	3:1	8.	8	28	3:7
3.	3	3	3:2	9.	9	36	3:8
4.	4	6	1:1	10.	10	45	1:3
5.	5	10	3:4	11.	11	55	3:10
6.	6	15	3:5	12.	12	66	3:11

Quilting Bee *(page 45-47)*

Task: Sample quilt blocks and spreadsheet data are shown.

The finished quilt has 56 quilt blocks, each 12" x 12", arranged in 8 rows of 7 blocks. The area of the border is 736 sq. in.; the total quilt area is 8800 sq. in. or 6.79 sq. yds. Further details are in the "gems" spreadsheet.

Spring

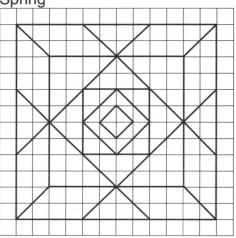

Steps

Gems

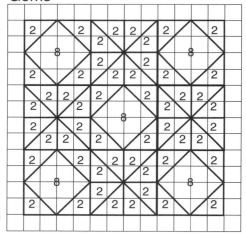

Gems Data Chart

shape	area-sq. inch	# in quilt blk.	# in quilt	total area
square	8	5	280	2240
sm. triangle	2	52	2912	5824
			total quilt blk. ar.	8064
			2" border (in.²)	736
			total quilt area	8800
			area (sq. yd)	6.79

Performance Assessments *(page 48):*

Assessment 1: 25.2; 24; 24; 22. The trapezoid and right triangle cover the same area as a 6 x 4 rectangle.

Assessment 2: 116 square units. Subdivisions will vary. Example: a trapezoid and a scalene triangle: $(.5)(9 + 13)(7) + (.5)(13)(6) = 77 + 39 = 116$ square units. Another possible subdivision: a 9 x 7 rectangle and three right triangles: Areas $= 63 + 14 + 27 + 12 = 116$.

FAST 4!

1 Estimate the area of the state of Nevada in square miles using the given dimensions.

Check your estimate with an atlas.

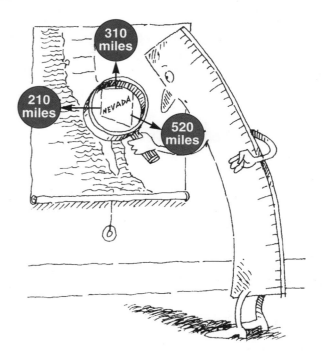

2 Plot and connect the ordered pairs (2,2), (7,2), (9,5), and (2,5) on a coordinate plane.

Name the figure you formed, and find its area.

3 The difference between the perimeters of these two triangles is 28 cm.

What is the area of each?

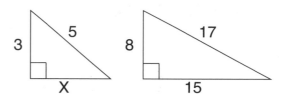

4 The diameter of the circle is 8 cm.

Find the area of the shaded region in the square.

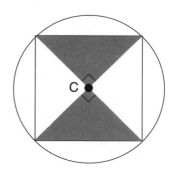

AREAS OF POLYGONS

THE TANGRAM PUZZLE

The seven-piece **tangram puzzle** originated in China. It includes five triangles (two small, one medium, and two large), a square, and a parallelogram. The story of Tan is often used to introduce tangrams.

Tan was the son of a Chinese nobleman. Tan's father gave him a beautiful red tile for his birthday, but as Tan played, it dropped from his hands and broke — into the seven geometric shapes of the puzzle! The wise man of the village asked Tan to study the pieces and tell him more about them, so he might perhaps be able to advise how to piece the shapes together again into the square tile shape.

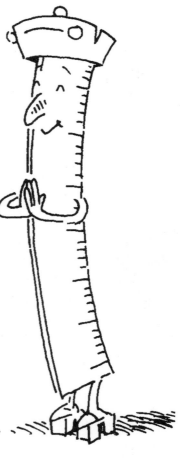

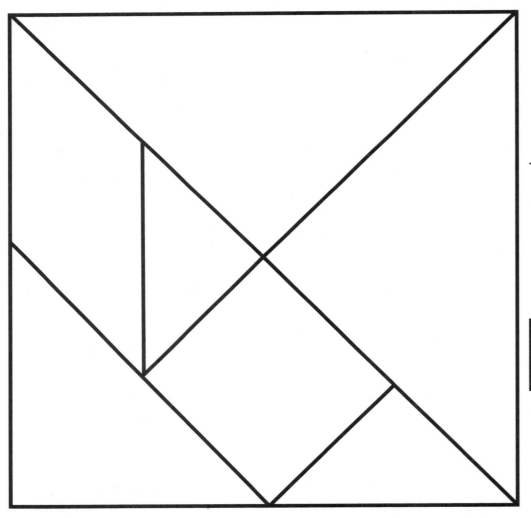

Cut out the seven tangram pieces along the solid lines.

AREAS OF OTHER POLYGONS

Triangle Area

What is the area of your "work of art"?

AREA ART

Triangles from the Tangram Puzzle, cm grid paper, lightweight cardboard (old cereal boxes or file folders), glue, scissors, markers or crayons, protractor, and a calculator.

WARM-UP

Trace the tangram triangles onto cm grid paper. Make templates by cutting out the grid paper triangles and gluing them to lightweight cardboard. Find the area of each triangle using two different methods. Explain to your classmates which method you prefer — and why — or write this up in your Math Log.

TASK

Create an original picture of an animal, object, or pattern using only tangram triangles. Trace the triangles from the templates. Find the total area covered by your picture in square cm. Show all calculations on your drawing paper, and write an explanatory paragraph explaining your strategy for calculating the area.

Ideas

- Use a minimum of ten triangles.
- Use each type of triangle at least once.
- Each triangle must touch at least one other triangle.
- Include a title for your drawing.
- Add color and details with markers or crayons.

NEXT STEPS...

Create a picture using all seven tangram pieces. Use a method to calculate the total area of your picture. Cut tangram pieces from brightly colored paper, and glue on construction paper using your design. Another time, trace and color your picture on a shirt using fabric crayons. Set with an iron.

AREAS OF OTHER POLYGONS

A coordinate game of logic

CAPTURE THE TANKS

For each player: Quadrant I-cm grids (battlefield) from page 39, 2 dice, a large book or folder, and a writing utensil.

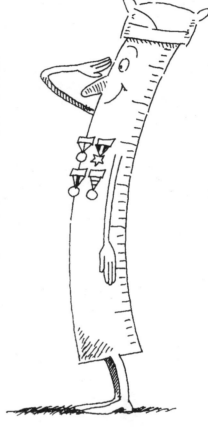

Game Rules

- Each player draws five triangle tanks on their battlefield before starting play:
  - Roll two dice to get the base and height of each triangle tank.
  - Sketch each tank onto a coordinate battlefield grid.
- Stand a book up to keep your opponent from seeing your battle positions.
- To play, take turns calling out an ordered pair for a battle position strike.
- Your opponent checks his/her own battlefield and declares:
  - Capture (if your call is within the area of a triangle tank),
  - Near Miss (if your call is within one square unit of a tank), or
  - Miss.
- If a player's tank is hit, the player must erase it and give the base and height to the opponent, who adds it to his/her fleet in any battlefield location desired.
- Continue playing for a set amount of time.
- When the battle time is over, players register all tanks on their tally sheets, recording dimensions and area of every tank in their possession.

Player 1:_____   Player 2:_____

Tank	Dimensions (cm)	Area (sq. cm)	Tank	Dimensions (cm)	Area (sq. cm)
		Total			Total

AREAS OF OTHER POLYGONS

Triangle Areas

A coordinate game of logic

CAPTURE THE TANKS

Battlefield 1

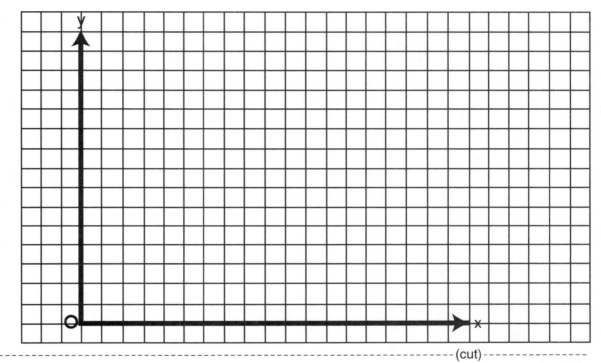

--- (cut) ---------------------

Battlefield 2

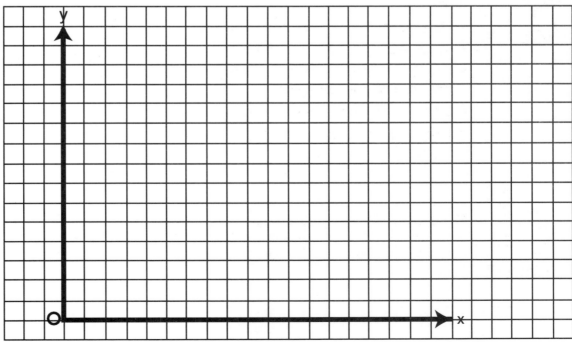

AREAS OF OTHER POLYGONS

measure and solve

Can you make polygons and measure their area?

PATTERN BLOCK POLYGONS

Pattern blocks, newsprint, metric ruler, pencil, calculator, and markers.

WARM-UP

- Use pattern blocks to construct a pentagon, hexagon, heptagon, octagon, nonagon, and decagon. Show sketches of your work in your Math Log. Label each figure and decide if it is **regular** or **irregular**.

- Discuss real-life words having the same prefixes as the shapes you sketched.

- Brainstorm a chart of common words containing prefixes penta-, hexa-, octa- and deca-. Where did these prefixes originate?

TASK

Construct a pentagon, hexagon, heptagon, octagon, nonagon, and decagon using pattern blocks.

- Use more than one block per shape.

- Trace each polygon onto newsprint.

- Label the dimensions in centimeters, and figure out a strategy for calculating the area of each polygon.

- Show calculations next to each figure.

- Use a black marker to highlight your figures and add color to your work if you wish.

COMMUNICATE

Write an explanatory paragraph describing how you calculated the area of each figure. Compare strategies you used to calculate areas with those of your classmates. Were some approaches more efficient than others?

Can you find areas of rectangles, squares, parallelograms, and triangles?

AREA BINGO

For each pair of players: bingo boards, 24 game cards in a bag, game markers, and a calculator.

Game Rules
- Mix up the game cards in a bag.
- Take turns drawing a card.
- Discuss and agree upon a solution with your partner.
- Cover the answer on the game board with a game marker.
- Score BINGO when you cover a diagonal, row, or column.

Cut out these
GAME CARDS

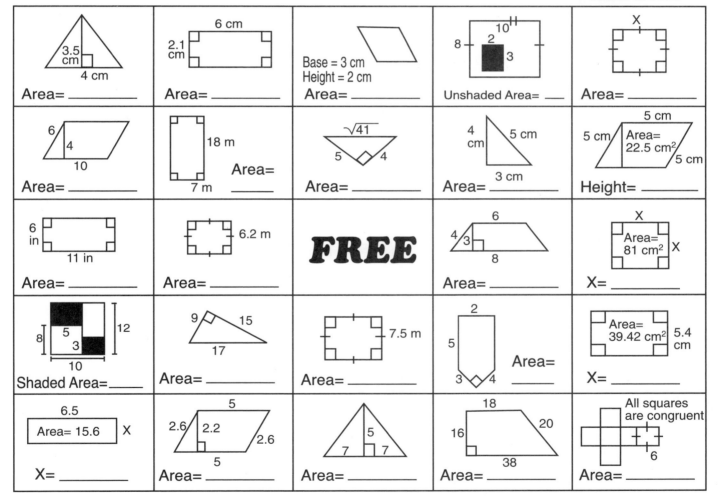

Row 1:
- 3.5 cm, 4 cm — Area= _____
- 6 cm, 2.1 cm — Area= _____
- Base = 3 cm, Height = 2 cm — Area= _____
- 10, 8, 2, 3 — Unshaded Area= _____
- X — Area= _____

Row 2:
- 6, 4, 10 — Area= _____
- 18 m, 7 m — Area= _____
- $\sqrt{41}$, 5, 4 — Area= _____
- 4 cm, 5 cm, 3 cm — Area= _____
- 5 cm, Area= 22.5 cm^2, 5 cm — Height= _____

Row 3:
- 6 in, 11 in — Area= _____
- 6.2 m — Area= _____
- **FREE**
- 6, 4, 3, 8 — Area= _____
- X, Area= 81 cm^2, X — X= _____

Row 4:
- 12, 8, 5, 3, 10 — Shaded Area= _____
- 9, 15, 17 — Area= _____
- 7.5 m — Area= _____
- 2, 5, 3, 4 — Area= _____
- Area= 39.42 cm^2, 5.4 cm — X= _____

Row 5:
- 6.5, Area= 15.6, X — X= _____
- 5, 2.6, 2.2, 2.6, 5 — Area= _____
- 5, 7, 7 — Area= _____
- 18, 16, 20, 38 — Area= _____
- All squares are congruent, 6 — Area= _____

Continue playing until your partner also gets BINGO. Or, choose to cover the board. Try creating your own area game with a blank BINGO board.

NEXT STEPS...

AREAS OF OTHER POLYGONS

Areas of Common Polygons

B	I	N	G	O
7.3 cm	6 cm²	56.25 m²	2.6 sq. units	4.5 cm
38.44 cm²	66 sq. in.	448 units²	6 cm²	11 units²
16 units²	10 square units	FREE	126 m²	21 units²
7 cm²	x^2 units	9 cm	67.5 units²	40 units²
12.6 cm²	74 units²	35 sq. units	216 units²	60 units²

B	I	N	G	O
38.44 cm²	66 sq in.	448 units²	6 cm²	12.6 cm²
126 m²	10 square units	16 units²	216 units²	11 units²
40 units²	67.5 units²	FREE	9 cm	56.25 m²
60 units²	35 square units	x^2 units	7.3 cm	2.6 sq. units
74 units²	7 cm²	21 units²	6 cm²	4.5 cm

AREAS OF OTHER POLYGONS

Can you compare the red and the blue areas of your trapezoids?

TRAPEZOID PATTERNS

Red trapezoid and blue rhombus pattern blocks, graphing calculator, and a writing untensil.

- Make a **trapezoid** with one blue and two red pattern blocks. Sketch it in your Math Log. Now figure out how many blue rhombi will cover the same area. Form these blues into a parallelogram having the same area as the trapezoid.

- Calculate the area of the parallelogram: A = b x h.

- Sketch the parallelogram in your Math Log and label its base, height, and area.

- Now go back and look at your trapezoid. Can you come up with a rule for calculating its area without changing its shape?

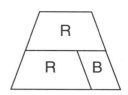

Look at these growing trapezoids. Make the first four figures with pattern blocks and sketch them in your Math Log.

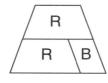

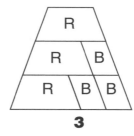

1   **2**   **3**   **4**

- Describe this pattern in writing. Figure out the pattern for both reds and blues.
- Share descriptions with your class.
- Decide: When will the number of blues be double the number of reds? Triple?
- Look at the areas covered by each color in each step of the pattern. When will the areas be equal? When will the area covered by blue be double the red area?

- Continue the pattern in your Math Log, recording letters R and B for the blocks in the pattern.

Step 5   Step 6   Step 7   Step 8

R
RB
RBB
RBBB
RBBBB

AREAS OF OTHER POLYGONS

Organize pattern data in the chart below for the first 12 steps of this pattern. Compare the area covered by reds to the area covered by blues in each trapezoid.

TASK

n=Step Number	# Red Trapezoids	# Blue Rhombi	Ratio of Areas R:B
1	1	0	1:0
2	2	1	3:1
3	3	3	
4			
5			
6			
7			
8			
9			
10			
11			
12			

- When does the blue area surpass the red area?
- At which step will the blue area be triple the red?
- Describe any patterns you see in the ratios of red area to blue.

COMMUNICATE

Make a scatter plot on a graphing calculator. Enter the ratios for blue and red areas in your x and y lists. Sketch your graph below, labeling the axes. Trace along the graph following the pattern out for at least 20 more steps. What do you notice? Next to your sketch, describe the graph.

SCATTER PLOT SKETCH

Description:

AREAS OF OTHER POLYGONS

Can you design an original quilt square?

QUILTING BEE

Graph paper, tangrams or pattern blocks, pencil, crayons or colored pencils, and a calculator.

WARM-UP

- Bring books showing quilt designs to class. Discuss the purposes behind quilting in early America.
- Look at a real quilt. What materials are needed to make one? How is a quilt put together?
- Look at different quilt blocks. Which shapes do you see most often in the patterns?
- Find a pattern using triangles. Which type of triangles are used?
- Sketch one square of your favorite pattern in your Math Log.
- Identify and label all geometric figures in the design.

A queen size quilt is 88 inches by 100 inches.

- If the quilt has a 2-inch border, how many 12 x 12 inch quilt blocks will be needed for the quilt? Explain how you know.
- Devise a strategy to calculate the area of the border.

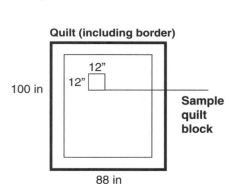

Quilt (including border)

100 in

12"
12"

Sample quilt block

88 in

TASK

Make a scale drawing of one 12-inch quilt block using tangram pieces or pattern blocks on $\frac{1}{2}$-inch graph paper (page 47). Figure out the area in square inches of each shape in the design. The information and ideas below should help.

Quilt Block Steps

- Using the scale $\frac{1}{2}$ square inch = 1 square inch, outline a 12" square quilt block.
- Arrange tangram shapes or pattern blocks into a pleasing design.
  You may reduce these shapes as needed to make the pieces fit.
- All geometric shapes must touch without gaps.
- Trace your design onto graph paper in pencil first, then color.
- Label the area on each shape.

AREAS OF OTHER POLTGONS

What is the total area of your quilt?

QUILTING BEE

AREA SPREADSHEET

Shape	Area of Shape :sq. in.	# in Each Quilt Block	# in Entire Quilt	Total Area
			Total Area of all quilt blocks (sq. in.)	
			2" Border (sq. in.)	
			Total Quilt Area (sq. in.)	
			Total Quilt Area (sq. yards)	

COMMUNICATE

Explain how you converted the number of square inches in your quilt to square yards. Compare your area to the total quilt area you figured from the 88" by 100" dimensions.
Do the areas match?

Try constructing your quilt square on Geometer's Sketchpad. Then copy and translate your square to create the larger quilt. Color the design with colored pencil.

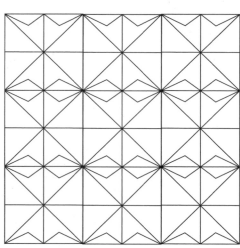

Quilt block designed on Geometer's Sketchpad

AREAS OF OTHER POLYGONS

QUILT SQUARE DESIGN

Pattern Name:_____

Scale:_____

AREAS OF OTHER POLYGONS

PERFORMANCE ASSESSMENTS 1-2

ASSESSMENT 1: **Rubric** 4   3   2   1   0

Which of the figures below covers the same area as a 4 x 6 rectangle?
Show your calculations and share your results in your Math Log.

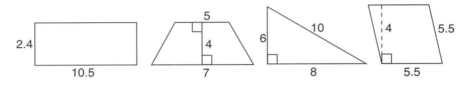

ASSESSMENT 2: **Rubric** 4   3   2   1   0

Find the area of the pentagon.

Show your calculations and explain your strategy
for finding the area in your Math Log.

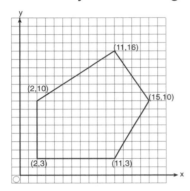

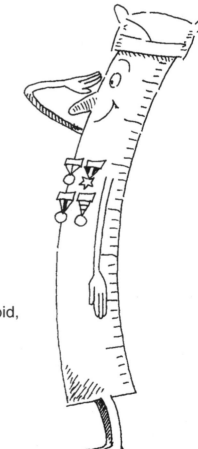

RUBRIC

4 Fully successful with tasks.

- Calculated areas of rectangles, parallelogram, triangles, trapezoid, and pentagon correctly.
- Can find area of polygons by subdividing them into polygons.

3 Substantially successful.

- Can find correct area of 4 out of the 5 polygons in the tasks.
- Showed evidence of a subdividing strategy in order to find area of the pentagon.

2 Partially successful.

- Found the correct area of 3 out of the 5 polygons in the tasks.
- Reteaching is necessary.

1 Engaged in the task

- Found the correct area of at least one polygon.

0 No attempt or non-mathematical response.

Section 4:
AREA OF CIRCLES

The following words are used throughout Section 4. An informal definition for these words is listed below. Teachers may want to highlight the words and review their meanings as they appear in the context of each lesson.

congruent: equal

sectors: parts (slices) of a circle

inscribed: a description for a polygon touching the points inside a circle

circumscribed: a description for a polygon touching the points outside of a circle

AREA OF CIRCLES

Fast 4! *(page 52)* - 3.14 used for π

- **Problem 1:** Area covered is 31,400 miles; circumference is 314 miles.
- **Problem 2:** About 1/16 of the lawn can be watered at one time.
- **Problem 3:** The volcanic crater Toba would be about 29.5 mi. across its center.

    Challenge: Toba's Crater is 9.3 times wider than Kilauea in Hawaii.
- **Problem 4:** Shape A: area of square (100 sq. units) – area of semi-circle (39.27 sq. units) = shaded area, 60.73 sq. units.

    Challenge: Shape B: area of circle (50.27 sq. units) – area of triangle (16 sq. units) = shaded area, 34.27 sq. units

Inner - Outer Areas *(page 55)*

Task: Younger students in particular may need to be encouraged to find the area of trapezoids or triangles that cover the hexagon.

Communicate: The area of the circle in each case is approximated by the average of the inner and the outer shapes, as the circle's area is always "in between" these inner and outer areas.

Reasoning: As students explore further, they should conclude that the average of the inner and the outer shapes remains a good estimate of the circle area, but that shapes with more sides provide better estimates of the circle area. (As the number of sides increases, the shape more closely resembles the circular shape.)

Next Steps: Area of outside square is $4r^2$; area of inside square is $2r^2$, so circle area is about $3r^2$.

Square a Circle *(page 55)*

Warm Up: A = 10 sq. units; B = 24 sq. units; C = 42 sq. units–each close to $3r^2$. Students may suggest values slightly more than $3r^2$.

Task: Area of a circle, A is about $3r^2$.

Next Steps: Half the circumference is πr, so A = $\pi r \times r$.

Chocolate Chip Cookies *(page 57)*

Warm Up: The area is 4 times larger. This pattern holds for other shapes as well.

Task: The diameter of a 16" cookie is double that of an 8" cookie, so it takes four 8" cookies to match what you get in one 16" cookie, or eight 8" cookies to match two 16" cookies.

Next Steps: Answers may vary slightly, but should be about 9 times more than the next smaller size. (Mega = 90¢, Jumbo = $8.10) This corresponds to the cookie area increase each time.

Close to 3000 Game *(page 59)*

As students become more skillful in using the relationships between the radius, diameter, circumference, and area measurements of given circles, they will use estimation strategies to get "close to" the targeted 3000 sq. units.

Communication: All students learn when they share their strategy and thinking. Use the questions to stimulate discussion.

Performance Assessments *(pages 60-61)*

Assessment 1: The area of the sheet metal is 162 sq. inches. Each circle has d = 3 or r = 1.5, so each has A ≈ 7.07 sq. inches. So the 18 circles have a total area of about 127.17 sq. inches, making a waste of approximately 34.83 sq. in.

Challenge: When the circles are nested, the length b can be determined by using the Pythagorean Theorem (≈ 2.6 in.). So x ≈ 1.1 inches, and the dimensions of the rectangle are 15 and ≈ 10.4. If sheet metal is available in these dimensions, its area is about 156 sq. inches, and the waste would be less (≈ 28.83 sq. inches) when the 18 circles are cut.

Assessment 2: Area of the first circle: 144π;
Area sum of the 4 circles: 4(36π) = 144π;
Area sum of the 9 circles: 9(16π) = 144π.
(This pattern continues.) Students may deduce that the sum of the areas of the inscribed circles is always 144π rather than carrying out the process 8 more times.

Assessment 3: Option 1 gives Fido more space to roam, as the sketches show.

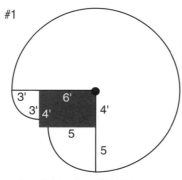

#1

.75 x 3.14 x 81 = 190.755 sq'
.25 x 3.14 x 25 = 19.625 sq'
.25 x 3.14 x 9 = 7.065 sq'
Roam space = 217.445 sq'

Option #1 gives Fido more space to roam

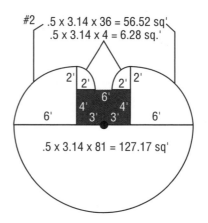

#2   .5 x 3.14 x 36 = 56.52 sq'
.5 x 3.14 x 4 = 6.28 sq.'

.5 x 3.14 x 81 = 127.17 sq'

56.52 + 6.28 + 127.17 = 189.97 sq'

FAST 4!

1 A typical hurricane might be about 200 miles in diameter. What area would a hurricane this size cover? What would be the circumference of a hurricane this size?

Challenge: About how much of your state would a hurricane this size cover? Are hurricanes likely in your state?

2 If the spray from a lawn sprinkler makes a circle 10 feet in radius, about how much of a circular lawn with a radius of 40 feet can be watered at one time?

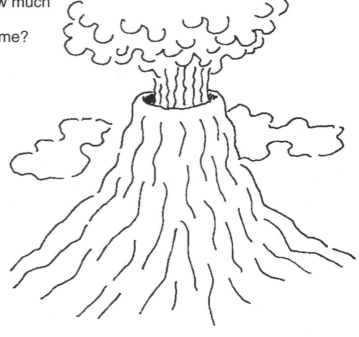

3 Toba, in Indonesia, is the largest volcanic crater in the world. It covers 685 sq. miles. If this crater were circular, about how many miles across the center would it be?

Challenge: About how many times bigger across is Toba's crater than the Kilauea crater in Hawaii, one of the most active volcanos in the world, which covers about 5 sq. miles?

4 Find the area of the shaded region in A.

Challenge: Find the area of the shaded region in B.

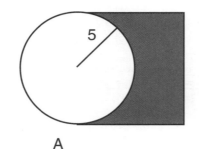

A

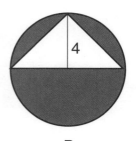

B

Approximating Circle Areas

How closely can you approximate the area of a circle?

INNER - OUTER AREAS

For each pair of students: square, triangle, and hexagon pattern blocks; compass, millimeter ruler, and a writing utensil.

Trace an orange pattern block square, then use a compass to inscribe the square in a circle. Next, circumscribe a larger square around the circle.

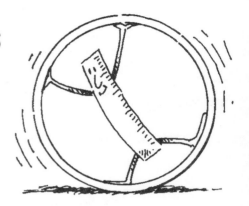

Find the areas to the nearest mm.of the:

- **inscribed** square
- **circumscribed** square.

Use this information to estimate the area of the circle.Write your estimate and your reasoning in your Math Log.

> **REMINDER:**
> A polygon that touches as many points of the circle as possible is:
> - **inscribed,** if inside the circle
> - **circumscribed,** if outside it.

Repeat the Warm-Up using yellow and green pattern blocks.

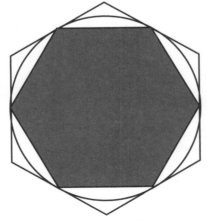

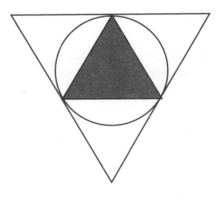

- In each case, how does the average of the inner and the outer shapes compare with the area of the circle?
- Write your conclusion(s) in your Math Log, and be prepared to share your thinking with the class.

AREA OF CIRCLES

Approximating Circle Areas

REASONING

Suppose you were to start with three circles the **same size,** and calculate the inscribed/circumscribed areas of:

- squares in the first circle;
- equilateral triangles in the second circle;
- regular hexagons in the third circle.

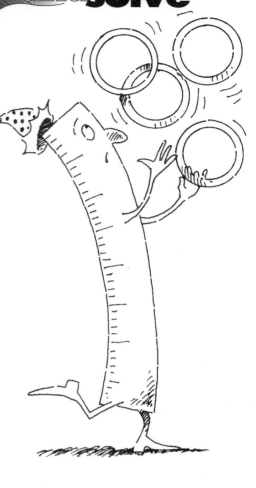

How do you **predict** your results would compare with your previous conclusions during Warm-Up and the Task? (That is, how do you think your results would be *like* what you previously wrote? How would they *differ*?)

Experiment as needed to confirm or modify your prediction(s).

NEXT STEPS...

How many r²s?

Study the inscribed and circumscribed squares to the right.

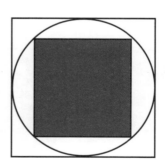

- Focus on the square on the outside of the circle and suggest a value for ☐ :

  Area of the circumscribed square = ☐ r^2

- Focus on the square on the inside of the circle and suggest a value for ☐ :

  Area of the inscribed square = ☐ r^2

Use this information to approximate the area of the circle.

Area of a circle, A is about ☐ r^2

AREA OF CIRCLES

How closely can you approximate the area of a circle?

SQUARE A CIRCLE

Scissors, paper, and a scientific or graphing calculator.

- Study the circles with centers at A, B, and C – with radius measures of 2, 3, and 4 units, respectively.
- Approximate the area of each shaded quarter-circle (counting each part-square as half a square).
- Estimate the area of circles A, B, and C.

Estimates of Circle Areas

A:_____ sq. units   B:_____ sq. units   C:_____ sq. units

For each circle, decide whether the area you estimated is closer to $3r^2$ or to $4r^2$.

Suggest a value for ☐ :

When calculating the area of a circle, A is about ☐ r^2

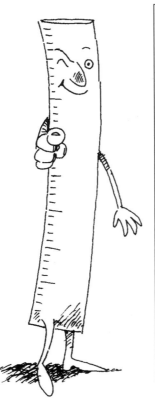

IT'S A FACT!

- The ancient Hebrews thought ☐ should be 3.
- Around 1650 B.C., the ancient Egyptians decided that ☐ was a little more than 3: $3^{13}/_{81}$!
- About 225 B.C., Archimedes, a Greek mathematician, basically agreed. He thought that ☐ was between $3^1/_7$ and $3^{10}/_{71}$.
- In 1707, an Englishman named William James was the first to use the symbol π (pi) for this number.
- Since then, computers have calculated the value of π to many, many decimal places!
- You can calculate the Earth's circumference within a fraction of an inch by using π to only 10 decimal places.
- Mathematicians have agreed that 3.14 or $^{22}/_7$ are close enough to π to make pretty accurate calculations.

π is a little more than 3

π = 3.14159265358979323846264338327950288419716939937511...

AREA OF CIRCLES

How can you find the formula for the area of a circle?

CIRCLES TO PARALLELOGRAMS

NEXT STEPS...

- Divide a paper circle into congruent sectors as shown, and rearrange them to form a "wavy" parallelogram.

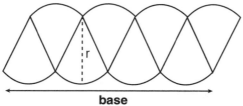

base
(half the circumference)

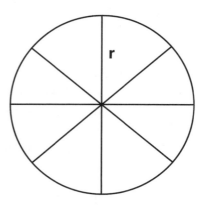

- Since C = 2πr, then half this, the base, is _____.

- If you were to repeat this process, making thinner and thinner congruent segments, the "wavy" parallelogram would get less wavy and look more and more like a real parallelogram. That is why a circle's area can be approximated by the area of the parallelogram, or

$$A = \underline{\hspace{2cm}} \times r$$

base x height

Did you find: **Area of a circle, $A = \pi r^2$?**

COMMUNICATE

Write a letter to a student in another class. Use sketches to help this student understand why the formula for finding the area of a circle makes a lot of sense. Be convincing!

Area of a circle, $A = \pi r^2$

Experiment with a scientific or graphing calculator that shows π. Find how many decimal places the calculator can show for π.

Can you get the most "cookie" for your money?

CHOCOLATE CHIP COOKIES

Pattern block squares and equilateral triangles
(21 of each per group of students).

Problem

The local cookie store advertised a special on its chocolate chip line.
A class of 25 students has a budget of $15 to buy cookies for a party.
How many of each size should they buy to get the most for their money?

Regular Cookie

4 in

45¢

8" Cookie

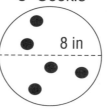

8 in

$1.50

16" Cookie

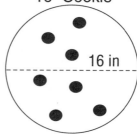

16 in

$6.00

WARM-UP

- Discuss with classmates which size cookie you think is the best buy.

- What if the cookies were square instead of round?

- Use pattern blocks to form 3 squares so that the side of the second is double the first, and the side of the third is double the second.

(Not actual size)

TASK

- How does the area increase each time you double the length of a side?

- Does the same pattern hold if you make larger cookies shaped like equilateral triangles?

Within your group, discuss the similarity of the square and triangle arrangements to the cookies on sale. Then explore the cookie problem above and make your recommendation.

AREA OF CIRCLES

REASONING

- Explain what area has to do with costing out how much "cookie" you get for your money.

- About how many regular cookies does it take to match the amount of "cookie" in an 8-inch cookie? In a 16-inch pizza cookie?

- About how many 8-inch cookies match the "cookie" in a 16-inch cookie?

- As the cookie size increases, study the increase in diameter sizes. How do they compare to the related area increases? In other words, when the diameter doubles, what happens to the area?

NEXT STEPS...

Suppose the manager of the cookie store wants to try a new line of cookies. The sizes are shown below.

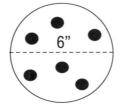

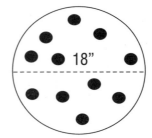

2" Mini
Cookie

6" Mega
Cookie

18" Jumbo
Cookie

The introductory sale price for the Mini is 10¢. To be consistent with this price, what should he charge for the other sizes? Explain your reasoning.

Instead of pattern blocks, use the Geometer's Sketchpad, if available, to draw and find the areas of the cookie shapes.

AREA OF CIRCLES

Circle Areas

A game for two 2-person teams

CLOSE TO 3000 GAME

Paper clip for spinner dial, calculator, and a writing utensil.

Objective: Notice how the radius, diameter, and circumference affect area.

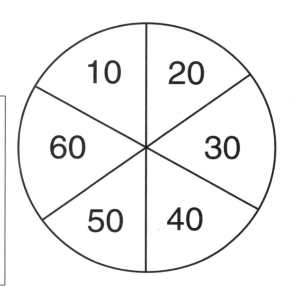

GAME RULES

- Spin. Decide whether you will use the number as a radius, a diameter, or circumference.

- Fill in your choice and find the area of a CIRCLE with that measure.

- Keep a running total of areas calculated. Try to get close to 3000 sq. units in 5 spins.

GAME SCORE SHEET

| Spin Number (ONE Choice Only) | | | Area of Circle With This Measure (sq. units to nearest 10th) | Running TOTAL of Areas (sq. units to nearest 10th) |
radius	diameter	circumference		
			TOTAL of Areas:	sq. units

COMMUNICATE

- How did you find the area when you only knew the circumference?
- What strategies did you find to help you get close to 3000 in 5 spins?
- Was any strategy more useful than others?
- Were any spin numbers more desirable than others?

AREA OF CIRCLES

PERFORMANCE ASSESSMENT 1

ASSESSMENT 1: Rubric 4   3   2   1   0

18 discs are to be cut out of
a sheet of metal measuring
18 in. x 9 in.

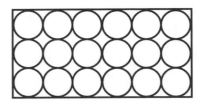

How much metal is wasted?
Show your calculations,
and explain your reasoning
in your Math Log.

Challenge: Would there ever be any advantage
to staggering the 18 discs as shown?

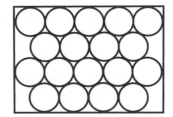

Show your calculations, and explain your reasoning in your Math Log.

RUBRIC

4 Fully applies area concepts in a problem
situation, with complete explanations.

3 Substantially applies area concepts in
a problem situation, with adequate
explanations.

2 Partially applies area concepts in a
problem situation with written
explanations.

1 Attempts to apply and explain area
concepts in a problem situation.

0 No attempt or
non-mathematical response.

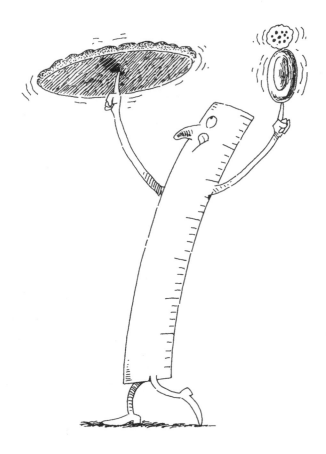

AREA OF CIRCLES

PERFORMANCE ASSESSMENT 2-3

ASSESSMENT 2: **Rubric 4 3 2 1 0**

Keep π in each answer rather than approximating it by 3.14 or 3¹/₇.

What is the area of the circle?_____

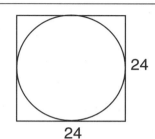

Subdivide this same-size square different ways.
Each time, find the sum of the areas of the inscribed circles.

Square ÷ 4

Area sum is _____ sq.units

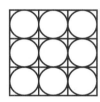

Square ÷ 9

Area sum is _____ sq.units

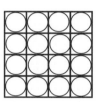

Square ÷16

Area sum is _____ sq.units

- If you continue using consecutive square numbers as the basis for subdividing as you repeat this process 8 more times, what will the sum of the areas of the circles be the last time you carry out this process?

- Show your calculations and justify your conclusion in your Math Log.

ASSESSMENT 3: **Rubric 4 3 2 1 0**

The Wilcox family has a new dog Fido. Since they have no fenced area in their yard, they plan to tie Fido on a 9-foot leash attached to a metal stake in the ground when he is outside. Where should they place the stake to give the dog the most space to roam?

- OPTION 1: At the corner of their 4 ft. by 6 ft. shed.

- OPTION 2: In the center of the 6-ft. side of the shed.

Show your calculations, sketches, and justify your conclusion in your Math Log.

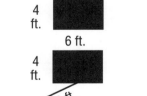

RUBRIC

4 Fully applies area concepts in a problem situation, with complete explanations.

3 Substantially applies area concepts in a problem situation, with adequate explanations.

2 Partially applies area concepts in a problem situation with written explanations.

1 Attempted to apply and explain area concepts in a problem situation.

0 No attempt or non-mathematical response.

AREA OF CIRCLES

1/4" Grid Paper

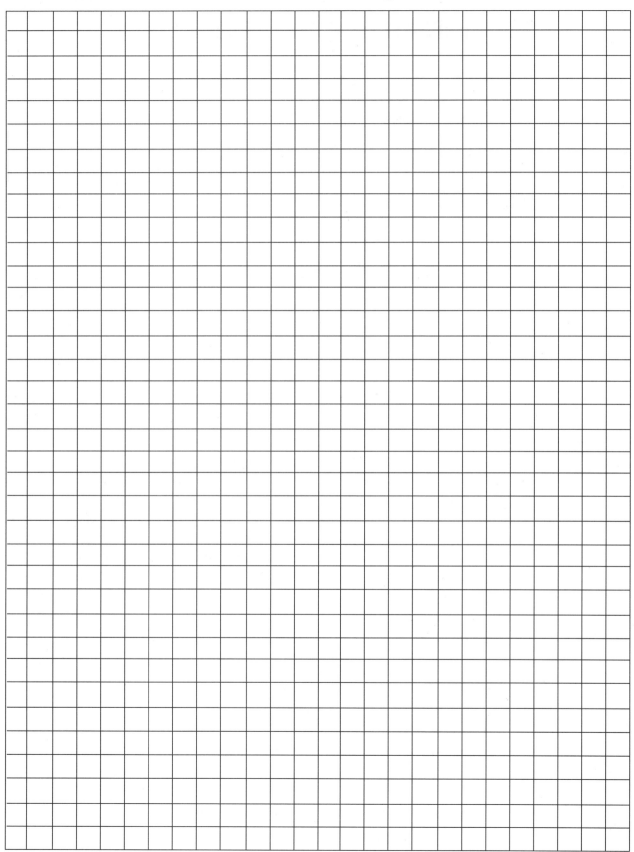

AREA OF CIRCLES

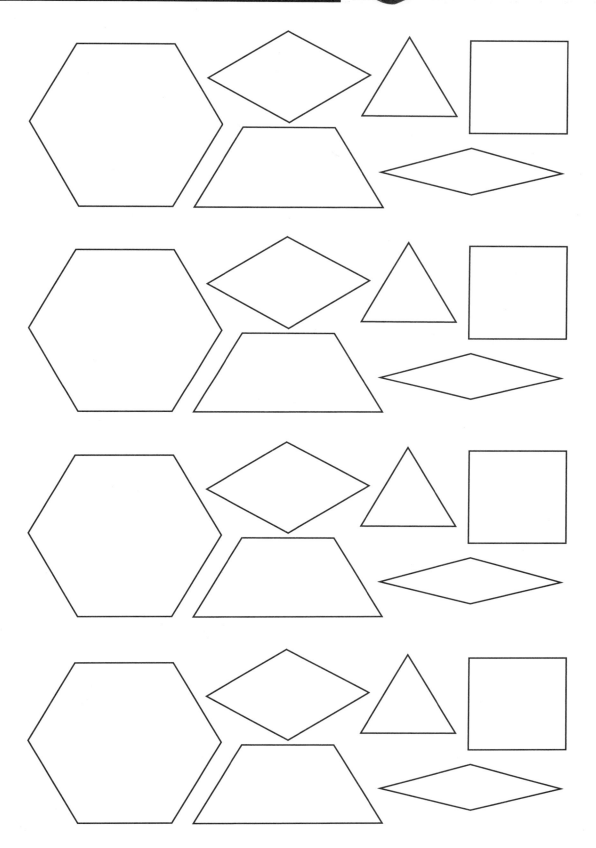

AREA OF CIRCLES

Color Tiles Blackline Master